The Adventures of Christopher Thornbuckle

Journey to the Stars

Library of Congress Control Number: 2026911695

Published by Discovery Trail Press™
An imprint of Adams Creative Enterprises

This is a work of historical fiction. While inspired by real events, locations, and historical periods, characters and events are fictionalized for storytelling and educational purposes.

Written by **Chris Adams**
Illustrated by **Chris Adams**
Edited by **Evan Adams and Michelle Adams**

First Edition

ISBN: 979-8-9951760-6-0

Note to the Reader

History is made of places, names, and dates. It is also made of people. People who lived long ago, just like you, made choices that helped shape our world.

When Christopher Thornbuckle travels to real places in time, he listens and learns from what he sees.

As you read this adventure to your preschooler, have them imagine standing beside him.

They will enjoy the book more when they become part of the adventure.

Let their imagination and creativity take over.

How to Read the Book

This is not a story to rush through. It is a story to read slowly and enjoy.

Each illustration helps create a visual experience while reading.

Engage your preschooler as they learn. Have them answer questions. They may ask some too.

Take your time and read closely. There is always something to discover.

How Christopher Came to the New World

Christopher Thornbuckle lives with his family on a small farm in the New World. Their journey began far away, in a city called Leiden in the Netherlands.

Christopher's parents had left England to find a place where they could live and worship freely. Life in Leiden was peaceful, but it was not always easy. Work was hard to find, and many families worried about the future.

In 1620, they sailed on a ship called the Mayflower. After many weeks at sea, they reached a new land.

They worked hard to build a home, start a farm, and make new friends.

Not far from their farm, deep in the trees, Christopher found something special. It was a glowing, magical passageway.

The Magical Passageway

Hidden in the woods near Christopher's farm is a special passageway.

When it is glowing, he can travel through time.

Christopher cannot make the passageway appear when he wants it to. When it appears, he has to be ready to go through it for his next adventure.

When Christopher returns back from the passageway, it is as if no time had past in his world. No one ever knows he is gone.

Welcome to *The Adventures of Christopher Thornbuckle*

Science for Preschoolers

The Adventures of Christopher Thornbuckle

PART I

Journey to the Stars

Christopher stepped outside. The sky was filled with bright lights.

He saw many tiny lights shining all around.

The next day he asked questions about the sky with his mom and dad.

"Christopher, the sun is very important. It gives us heat to keep us warm and helps things grow."

What grows under the sun?

We are going to sleep under the sky tonight by the ocean. I want to show you the stars and the moon.

Christopher was excited to see the sky by the ocean.

When I was young, your grandfather taught me all about the stars, the sun, and the moon. He loved the sky.

The next morning, they got up to watch a beautiful sunrise.

What ocean does the sun rise from in the east?

In the early evening, while looking up at the stars, they saw a shooting star.

The sky became darker. The moonlight comes from the sun shining on it.

Tonight the moon looks different.

What causes the moon to look different?

The next day at home, Christopher was telling his friends what he had learned.

What shines on the moon to make it glow?

On an autumn evening with the leaves turning colors, a harvest moon was in the sky.

What is a harvest moon?

Christopher asked his father if they could camp outside of their home to learn more about the stars.

"Look here, Christopher. That is the Big Dipper,"
his father said.

That's the North Star. It will always help us find our way home.

The next evening, a glow appeared at the edge of the forest. Christopher wanted to get closer to see it.

Christopher, being brave, decided to walk into the glowing light.

He first stuck his hand through it and noticed something different.

As he walked through the passageway, he was floating in space with a strange suit and helmet on his head. He was scared.

What do they call people who go to space?

He now understood the magical passageway
could take him to a different place.

He was still scared when he went back, so he
quickly went back home.

A few days past. Now, being brave, he entered the passageway with a smile. The planet was yellow and orange in color.

This planet was colorful with lots of blue and green. It looked like home.

Learning on how to float in space, Christopher started doing tricks while looking at a red planet.

Another planet was very large.

He was now floating above a planet with large rings around it.

He looked back with a smile seeing his first ice planet with rings around it.

Seeing another dark blue ice planet, he knew he was very far away from home. He really liked the color.

Why are these planets called "Ice Planets?"

A very small planet appeared. It is the farthest planet from the sun.

What is the name of this mini-planet?

Our Solar System
Sun
Mercury
Venus
Earth
Mars
Jupiter
Saturn
Uranus
Neptune
Pluto (Mini Planet)
Have fun saying the planets out loud!

“Father, don’t you see the glowing passageway?” “No, I do not, Christopher. You have such an imagination,” he said to his young explorer.

Christopher had seen many planets and learned about the stars in the sky. He thought, “What is beyond the stars and planets?”

Only the magical passageway knows the answer!

Discover more history beyond the pages of this book.
Visit our Learning Resource Center for activities, timelines, and free teaching guides:

discoverytrailpressbooks.com

DISCOVERY TRAIL PRESS

www.ingramcontent.com/pod-product-compliance
Lightning Source LLC
LaVergne TN
LVHW070626061126
851857LV00002B/18
9798995176060